The Night Book

Exploring Nature after Dark with Activities, Experiments and Information

written by Pamela Hickman

illustrated by Suzanne Mogensen

Kids Can Press

In this book you'll find activities to do both during the day and at night to discover more about the world after dark. The activities are labeled "Nighttime Activity" or "Daytime Activity." Remember always to take an adult along when you do a nighttime activity.

For my brother Michael

First U.S. edition 1999

Text copyright © 1996 by Pamela Hickman
Illustrations copyright © 1996 by Suzanne Mogensen

Published in Canada by
Kids Can Press Ltd.
29 Birch Avenue
Toronto, ON M4V 1E2

Published in the U.S. by
Kids Can Press Ltd.
85 River Rock Drive, Suite 202
Buffalo, NY 14207

Edited by Laurie Wark
Designed by Marie Bartholomew
Printed in Hong Kong by Wing King Tong Co. Ltd.

CMC 99 0 9 8 7 6 5 4 3 2
CMC PA 99 0 9 8 7 6 5 4 3 2

Canadian Cataloguing in Publication Data

Hickman, Pamela
    The night book : exploring nature after dark with activities, experiments and information

Includes index.
ISBN 1-55074-318-X (bound)
ISBN 1-55074-306-6 (pbk.)

1. Nocturnal animals — Juvenile literature.
2. Nature study — Juvenile literature.
3. Scientific recreations — Juvenile literature.
I. Mogensen, Suzanne. II. Title.

QL755.5.H53 1996   j591.5   C95-932688-X

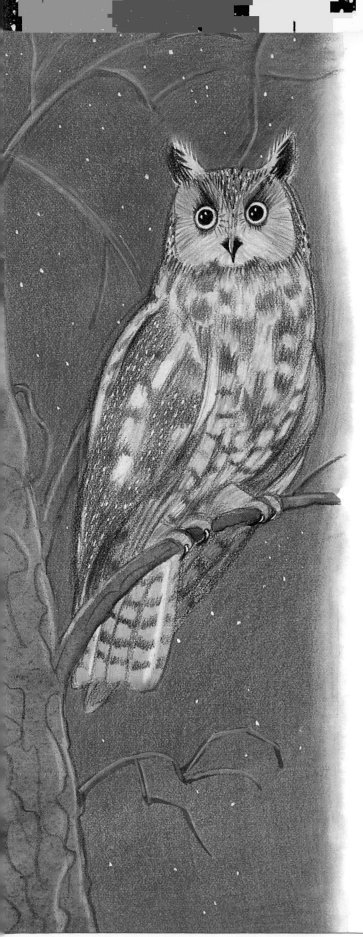

Contents

What is night?

What does the night mean to you? It may be a time when it gets dark and you go to bed, or when the stars come out. In your home, night may be a quiet time for rest, but in nature it is full of activity. This book introduces you to many fascinating animals that come out at night. You will also find out how plants attract nighttime visitors, and you'll meet some incredible creatures that glow in the dark. You can discover where the Sun goes at night, do some stargazing, and find out about the phases of the Moon. Have fun trying the nighttime and daytime activities, too. This book will wake you up to the amazing world after dark.

Day and night

You can find out why it gets dark every night by making a simple model of Earth and the Sun.

You'll need:
- a small ball
- a flashlight
- a sticker or tape

1. Place the sticker on the ball (or use a small piece of tape). This marks your home on Earth.

2. In a dark room, set the flashlight on a table and turn it on. The flashlight represents the Sun.

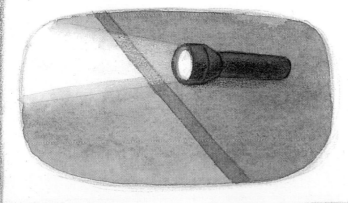

3. Hold the ball in front of the light so that the sticker side is lit up. It is daytime at your home.

4. Slowly turn the ball around in your hand until the sticker side is in the dark. It is nighttime at your home.

5. Turn the ball around again.

What's happening?
Earth is like a ball that spins around once every 24 hours. When the Sun shines on Earth, it lights up only half of the planet. While that half of Earth has day, the other half has night. Night is the time when the Sun isn't lighting the part of Earth where you live.

Moon watching

Nighttime Activity

Can you name a big rock in space that revolves around Earth once a month? The answer is the Moon. Spaceships have photographed it and astronauts have landed on it, but you don't need to ride in a rocket to see what the Moon looks like; all you need is a pair of binoculars. On a cloudless, moonlit night, lie on the ground or in a reclining lawn chair where you'll have a clear view of the Moon. Use pillows to prop up your arms so they don't get tired holding the binoculars.

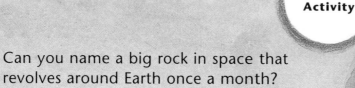

Here's what to look for:

• Find some large, dark areas on the Moon. These are huge sunken areas called seas, or maria. Unlike the seas on Earth, these don't have any water in them. You may see some of the biggest maria without binoculars.

• Look for round holes, or craters, in the brighter areas on the Moon's surface. These craters were formed when objects in space, such as chunks of rock, crashed into the Moon's surface.

• The Moon has mountains on it, just like Earth does. There are several large mountain ranges on the Moon that you may see with your binoculars.

Moon phases

Why do you sometimes see a Full Moon, and at other times you see only part of the Moon or no Moon at all? The light of the Moon comes from sunlight that is reflected, or bounced, from the Moon's surface. When the Moon is between Earth and the Sun, the sunlight hits the back of the Moon and we can't see it. This is called the New Moon. As the Moon revolves around Earth, a bit more sunlight shines on the front of the Moon each day, so we see more of it. Fifteen days later, it has traveled to the opposite side of Earth from the Sun. Here, the Sun shines directly on the front of the Moon and we see a Full Moon. Every day after that, the amount of Moon that we can see gets slightly smaller, until it disappears again and we have another New Moon. It is 29.5 days from one New Moon to another.

Full Moon

Last Quarter

First Quarter

New Moon

Some calendars show you the phases of the Moon each month. Check yours to see if it has these symbols.

Star light, star bright

We say that the stars come out at night, but actually stars are in the sky during the day, too. You can't see them because the light from our nearest star — the Sun — is so bright it blocks out the light from all the other stars. At night, when the Sun is not shining where you are, other stars can be seen.

On a dark, moonless night you may see about 2000 stars if you are out in the country, away from city lights. But there are many more stars in the sky than that. In fact, scientists think that there may be as many stars in the sky as there are grains of sand on Earth!

Watching stars is a terrific hobby. Even in the city you can see some of the brightest stars and constellations, or groups of stars.

A pair of binoculars is helpful for star watching. You may notice that some stars are brighter than others. This is because stars are different sizes and different distances away from Earth. Look for a hazy, riverlike pattern of stars flowing across the sky. This is called the Milky Way. Some of the stars you see may actually be planets instead, such as Venus and Jupiter.

Shooting stars and satellites

When you're star watching, you may see bright streaks of light pass quickly through the sky. These meteors, sometimes called shooting stars, are tiny bits of stone or metal burning up as they pass into Earth's atmosphere. August 10 to 12 is one of the best times to watch for meteors, during the Perseid meteor showers. At this time you may see up to 50 meteors an hour, all moving away from the constellation Perseus. The best time to see meteors is between midnight and dawn.

If you notice a starlike light traveling in a steady path across the sky, you've likely spotted a satellite that was sent up into space by scientists. Satellites are used for lots of different jobs, including taking photographs in space and sending out television and radio signals. The light you see is the reflection of sunlight on the metal body of the satellite.

When you wish upon a star, do you ever think about what it is made of? All stars are huge balls of hot, glowing gases. Stars come in different colors, depending on how hot they are. Red are the coolest, yellow are medium hot and white or blue-white stars are the hottest. What colors can you see?

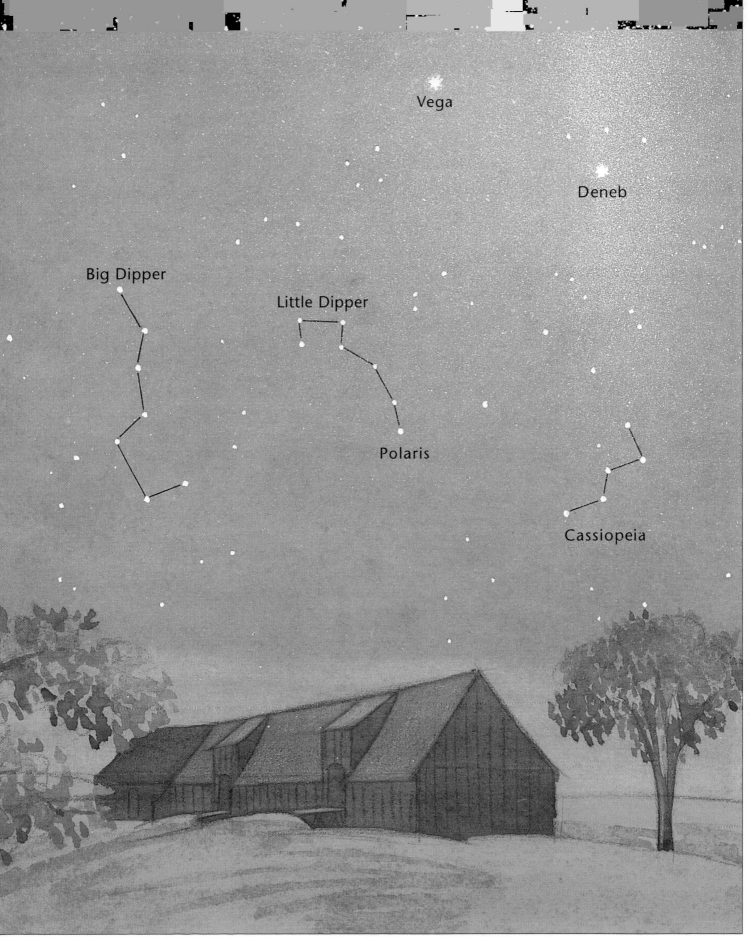

Vega

Deneb

Big Dipper

Little Dipper

Polaris

Cassiopeia

Stargazing

Nighttime Activity

Face northward and use this star map to look for constellations where you live. Turn the book so the current month is at the bottom of the map. On a clear, dark night, you should be able to see the constellations near the bottom and middle of the map. The Big Dipper, the Little Dipper, Cassiopeia and the Dragon are visible at any time of year. Each night the stars rotate around the very bright star Polaris, also called the North Star.

April
March
May
February
June
January
July
December
August
November
September
October

Virgo
Leo
Canis minor
Canis major
Ursa major
Bootes
Orion
Ursa minor
Draco
Scorpio
Polaris
Lyre
Twins
Cassiopeia
Cepheus
Cygnus
Aquila
Pegasus

Northern lights

If you live in the north, on a clear night you may see waves of spectacular colors moving and shimmering across the sky. The farther north you are, the more vivid the colors will be. What you're seeing are the northern lights. They are caused by tiny particles from the Sun entering Earth's atmosphere. The colors may be blue, pink, green, white, red and yellow. Toward the South Pole these colorful displays are called, not surprisingly, southern lights.

If a rocket took six months to travel to the Sun, it would take over 100 000 years to reach the next closest star.

Make a star scope

Daytime Activity

You can practice looking at different constellations during the day with this easy-to-make star scope.

You'll need:
- some cardboard tubes from toilet-paper or paper-towel rolls
- black construction paper
- a star map (see the map on page 10)
- a pencil
- a pin
- scissors
- tape

1. Stand the cardboard tube on a piece of black construction paper and trace around it. Cut out the circle.

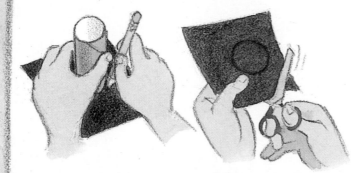

2. Choose one constellation from the star map and copy it onto your black circle, using pencil dots.

3. With the pin, poke a hole through each pencil dot, or star, in the constellation.

4. Tape the circle to the end of the cardboard tube. Make sure the tape doesn't cover any of your pinholes.

Naming constellations

5. Hold the open end of the tube up to your eye and point the other end at a window or light. You should be able to see the constellation shining through.

6. You can use several tubes and make a different constellation for each star scope.

7. Make different constellations with your friends. Take turns looking through the star scopes and guessing which constellations they show.

In ancient times people saw patterns in some groups of stars and gave them names. Here is the meaning of some common constellation names. Try making a star scope for some of these constellations.

Aquila	the eagle
Cygnus	the swan
Cepheus	the king of Ethiopia in Greek mythology
Cassiopeia	the wife of Cepheus in Greek mythology
Ursa Major	the great bear
Gemini	the twins
Canis Major	the great dog
Canis Minor	the lesser dog
Virgo	the maiden
Orion	the hunter
Pegasus	the winged horse
Scorpius	the scorpion
Perseus	hero
Polaris	North Star or polestar
Taurus	the bull
Corona Borealis	the northern crown
Bootes	the herdsman
Draco	the dragon

How animals sleep

Imagine sleeping upside down like a bat or taking 90-second power naps the way the blind Indus Dolphin does. If you were a kingfisher, you'd sleep on a bed of fish bones instead of your cozy mattress. Read on to find out more about bedtime in the animal world.

• Some bats wrap their wings around their bodies like blankets, and big-eared bats may tuck their ears under their wings for warmth.

• Many animals curl up in a furry ball with their nose and tail touching to keep warm. A fox's bushy tail makes a great pillow.

• Horses, elephants and other large animals can nap standing up, but they lie down for long sleeps.

• Virginia Opossums snooze for 18 hours a day, but giraffes need only 2 hours of sleep a day.

Do you know why a bird doesn't fall off its perch when it sleeps? When the bird's toes wrap around the perch, special muscles in each leg lock the toes into place so that they won't slip, even when the bird is asleep.

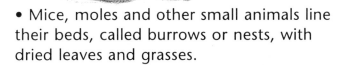

• Mice, moles and other small animals line their beds, called burrows or nests, with dried leaves and grasses.

• Fish and turtles head to the muddy bottom of a marsh or pond to sleep.

Sweet Dreams

If you watch a cat or dog sleep, you may see it twitching, growling, wagging its tail or moving its feet. That's because it is dreaming. Most mammals and birds dream. People sometimes move around or talk in their sleep when they are dreaming, too. Dreaming is a regular part of your sleep time, and everyone dreams. You dream at least four times every night even though you may not remember your dreams. When you are dreaming, your eyes move around beneath their lids. You can watch people sleeping and tell if they are dreaming by whether or not their eyes are moving.

Nocturnal animals

When you go to bed in the evening, many animals are just waking up. After spending the day resting, some animals go out to find food or mates during the night. Animals that are active at night are called nocturnal. Staying up at night might sound like fun to you, but it is a matter of survival for nocturnal animals. The dark helps to hide small animals from their enemies. Creatures that need to keep their bodies moist come out at night, too, when the air is damper and cooler. And hunting at night helps some predators avoid competing for food with daytime hunters.

• Both swallows and bats eat flying insects — swallows by day, bats by night.

• Otters aren't naturally active at night, but they become nocturnal when they live near people.

• Crayfish and other crustaceans shed their shells only at night, so they can hide from enemies while their new shells harden.

• Great Horned Owls and Red-tailed Hawks eat the same prey. The owls hunt at night and the hawks hunt during the day, so they can live in the same habitat without fighting over food.

• Mice come out at night when it's harder for foxes to hunt them.

• Earthworms, salamanders and slugs need to stay damp, so they avoid hot sunlight and come out at night.

Changes in the night

Insects usually change into adults, or emerge from their pupal case, at night, because it is the safest time for them. A new adult insect's skin is soft, and its body could easily dry out in the Sun. The insect also moves very slowly and can't fly for several hours after it emerges. If you take an early morning walk by a stream or marsh in summer, look on rocks and shoreline plants for the shed skins, or casts, of dragonfly and damselfly nymphs.

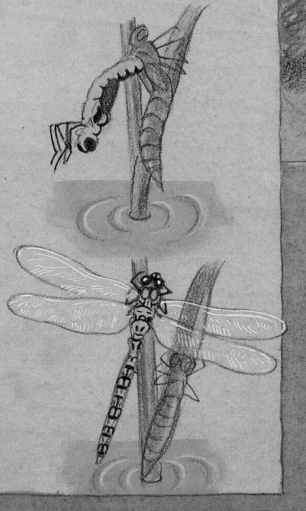

Night senses

If you tried walking outside in the dark, you'd probably trip and bump into things. Animals that are active at night are specially equipped with super senses to help them stay safe in the dark. On these pages you will discover how animals use their senses of hearing, sight, smell, taste and touch to find food and mates and to avoid danger in the dark.

Now hear this

Many nocturnal animals rely on their ears for survival. Barn Owls can hunt by hearing alone. Bats also use sound to find food and to avoid bumping into things while flying at night. They send out high-pitched squeaks and then listen for the echoes that bounce off nearby objects. This is called echolocation. Using echolocation, a bat can tell where an insect is, how big it is, how fast it is moving and in what direction. Some whales and porpoises also use echolocation to find prey in the underwater darkness.

Bring on the bats

You can get a closer look at bats and see echolocation in action with this simple trick. On a warm evening in summer or early fall, go with an adult to an open area with nearby trees. Pick up some wood chips or pebbles and carefully throw them into the air above you, one at a time. Bats in the area may discover the wood chips or pebbles with their echolocation and fly in to investigate.

Good vibrations

Most insects can't hear sounds like we do. Instead, they feel vibrations from the ground, caused by something moving nearby. Vibrations warn insects of approaching danger or food. Many birds, too, have tiny organs on their legs that pick up vibrations. This helps alert them to danger when they're sleeping.

Seeing in the dark

Most animals, including people, have two types of cells in their eyes, called rods and cones. Cones help to see detail and color, and rods are used for seeing in low light. Nocturnal animals have mostly rods in their eyes, so they can see well at night. Some bats, lizards and snakes have only rods and no cones. They are color-blind, and everything looks black and gray to them.

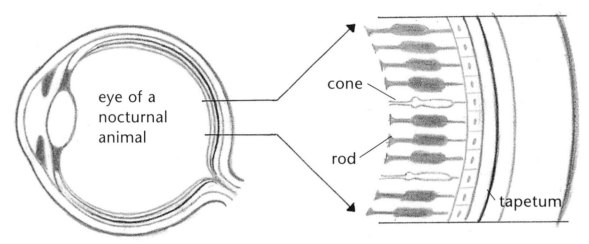

In this cross-section close-up of a nocturnal animal's eye, you can see that they have mostly rods and very few cones in their eyes. Human eyes have both rods and cones, with most of the rods toward the edges.

Test your night eyes

The next time you go out night watching, try looking directly at an object such as a star. Then look just to the side of the star. You can see the star more clearly when you don't look directly at it. This is because you have more rods at the sides of your eyes than anywhere else, so you can see best in the dark by looking to the side of an object.

Eye shine

Have you ever seen a cat's eyes glowing at night when a light shines on them? The glow comes from a mirrorlike layer at the back of the eye, called the tapetum. Make a simple model of a cat's eye to see how it works.

You'll need:
- black construction paper
- two plastic margarine or yogurt containers
- a small mirror (pocket-sized)
- tape
- a flashlight

1. Press a piece of black paper into each container so that the insides are completely covered.

2. Tape the mirror in the bottom of one container. The mirror works like the tapetum.

3. Put both containers on their sides in a very dark room or closet.

4. Shine the flashlight into the container without the mirror. Now shine it into the container with the mirror. What do you see?

What's happening?

When light enters a cat's eye, the light is bounced back, or reflected, through the eye by the tapetum. This increases the amount of light the cat sees, since the light is passing through its eye twice. This helps cats and other night animals see better in low light.

More common senses

Do you hold your hands out in front of you to feel your way around a dark room? Do you think you could find your friend in the dark by smelling for her? Some nocturnal animals have amazing senses to smell, taste and feel their way around at night.

• Most birds have a poor sense of smell, but sea birds such as fulmars, albatrosses and shearwaters use smell to find fish near the water's surface at night.

• Insects use their antennae to help them smell, taste and feel things. A male moth can find a female in the dark more than a kilometer (1/2 mile) away just by using his sense of smell.

• In dark waters, fish use smell to find other fish of their species and to tell males from females. Catfish, cod and carp have whiskerlike barbels growing around their mouths that can detect chemicals in the water. They drag their barbels along the bottom to help them feel, smell and taste what is down there.

• Raccoons use their keen sense of smell to find food in the dark.

• Deer and rabbits use their sense of smell to detect danger.

• The long, stiff hairs growing from its eyebrows, cheeks and mouth area help a cat feel its way around in the dark.

• A Starnose Mole's tentacles move around constantly, feeling in the dark and helping the mole find food.

• Slugs and snails use small pairs of tentacles to help feel their way through the dark.

• Snakes use their tongues for smelling. The tips of the forked tongue collect tiny scent particles from the air and bring them back into a special scent organ in the snake's mouth.

Nature's night-lights

Seeing in the dark is much easier when you have a flashlight. Some animals have built-in flashlights to help them see at night, find food, attract a mate or avoid enemies.

Creatures that can light up in the dark are called bioluminescent. The light is created by a chemical reaction, usually inside the animal's body. These living lights may be colored greenish blue, yellow or red, and some species make more than one color.

At night the ocean lights up with the glow of many creatures, including jellyfish, octopuses, clams, worms, shrimp, snails, fish, bacteria and algae. The angler fish has a wormlike growth dangling near its mouth that looks like a fishing lure. The "lure" lights up in the dark and attracts other creatures close enough for the angler fish to catch them. The Flashlight Fish of the Red Sea and Indian and Pacific oceans has a

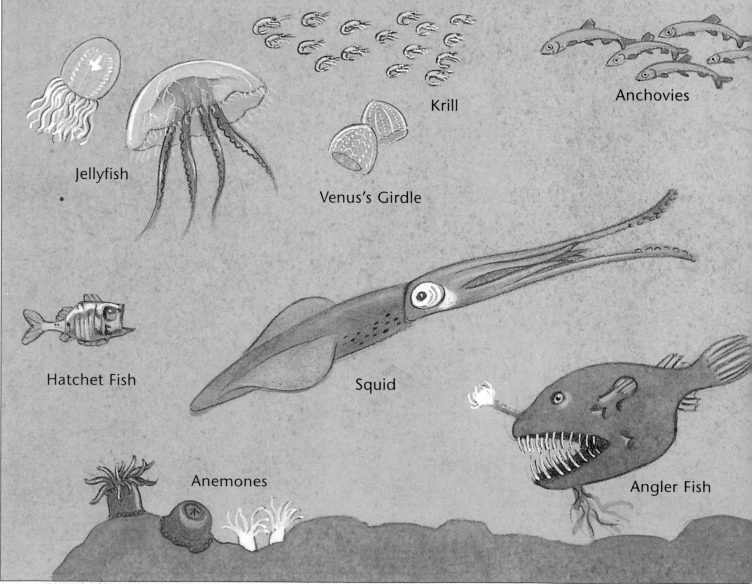

Krill

Anchovies

Jellyfish

Venus's Girdle

Hatchet Fish

Squid

Anemones

Angler Fish

patch of bacteria that glows under each of its eyes. The blue-green glow of the bacteria helps light up the dark ocean waters so the fish can find food. Special folds of skin under each eye, like eyelids, can cover the lights to help the fish avoid enemies.

You may find glow-in-the-dark creatures on land, too. Some millipedes, centipedes and earthworms produce their own light, and so do insects such as fireflies, glow-worms, ground beetles, click beetles, midges and springtails. Even some mushrooms have glowing parts, but scientists aren't sure why they glow.

Sea Gooseberry

Flashlight Fish

Firefly watching

The main reason why animals light up is to send messages to each other, especially when they are searching for a mate. You may have seen the flashing lights of fireflies on warm evenings in late spring or summer. Male fireflies fly around and send signals to the females that are resting on or near the ground. Each species of firefly has its own code, or pattern, of flashes, and a female responds to the code for her species. When a female flashes back, the male swoops down to mate with her.

If you see fireflies flashing, watch their light patterns closely. Can you see the signals of different species? Look for both male (flying) and female (resting) signals.

Birds at night

As it gets dark, most birds return to familiar roosts to sleep — in trees, on buildings or other high places. A nighttime roost can sometimes be a very crowded place. For instance, a single chimney may be home to hundreds of Chimney Swifts that roost there every evening. But not all birds go to bed at night.

Snipe

Kiwi bird

Whippoorwills, nighthawks and other members of the bird group known as goatsuckers spend their nights catching night-flying insects. These birds open their large mouths to catch insects while they fly. Whiskerlike feathers around their mouths help them trap food.

Many night birds such as woodcock and snipe feed on earthworms and other soil creatures. The kiwi bird of New Zealand also hunts for worms at night, but it can't see very well with its small eyes. To find worms, a kiwi pokes its beak in the ground to feel and smell for them. Most birds have their nostrils near their face, but a kiwi's nostrils are near the tip of its long beak, where it can sniff for food underground.

The Oilbirds of Trinidad and northern

Whippoorwill

Oilbird

South America are the only nocturnal birds that eat fruit. Their super sense of smell helps them find food in the dark. Like bats, Oilbirds also use echolocation to help them find their way in and out of the dark caves where they roost during the day.

Nighthawk watch

Ask an adult to join you for some nighttime birdwatching. Get some binoculars and head out to your backyard or neighborhood park. You may see the Common Nighthawk flying high in the air over a city or in open country after dusk on summer evenings. Listen for its nasal "peent" call, and watch for its long, pointed wings with their white patches. Nighthawks are amazing insect catchers. One bird was discovered with over 500 mosquitoes in its stomach, while another had eaten 2175 flying ants!

Owling

Seeing an owl is a rare and exciting event, since owls are mostly active at night when people are asleep. You may hear the thrilling, chilling sound of an owl calling at night. It calls to communicate with other owls and claim its territory. If you learn to imitate an owl's voice, you may get one to answer back to you. Look at this picture to discover how an owl is adapted for hunting in the dark.

Large, sensitive ears for hearing prey are hidden under the feathers at the sides of its saucer-shaped face.

Its large eyes let in lots of light and help the owl to see in the dark. Both eyes point forward, so the owl can see depth and details far away.

An owl's flight feathers are fringed so air passes through them silently and the bird can sneak up on its prey.

An owl's body is light in relation to the size of its wings, so it doesn't need to flap as much as other birds. This also helps it make less noise while hunting.

What do owls eat at night?

When an owl eats, it swallows its prey whole. It can't digest the bones, fur, claws, beaks and other tough bits, so its body forms tiny, dry "packages," called pellets, which the owl spits up. Scientists use owl pellets to help figure out what owls eat in different habitats and at different times of the year. You can find out what an owl eats by looking at its pellets, too. Pellets are usually found under an owl's daytime roost or nest or under a nighttime feeding perch. Many owls roost during the day and nest in large trees in wooded areas. Their night perch is often in a tree or on a fence post at the edge of a field.

You'll need:
- owl pellets
- a dish of warm water
- paper towels
- tweezers or two skewers

1. A small pellet can be dissected dry, but soak a large one in a dish of warm water for an hour or so.

2. Place the pellet on some paper towels. Use the tweezers or skewers to gently separate the bones, fur and other bits.

3. If you find feathers, you know the owl has eaten a bird. Tiny bones and bits of fur probably belong to a mouse, shrew or other small animal. You may also find insect parts such as wings or legs. Try to figure out what the owl has been eating.

Night fliers

You've probably noticed the crowds of insects flying around streetlights on summer nights. Many insects come out in the dark to avoid being eaten by daytime hunters such as birds. Have you ever wondered why insects are attracted to light? Night-flying insects use moonlight as a guide for flying in the dark. Since the Moon is so far away, insects can fly in a straight line by keeping the moonlight at the same angle to their eyes all the time. When the insects see the brighter, close-up light of a porch light or candle, they get confused and use the nearer light as a guide instead. Unfortunately, in order to keep the nearby light at the same angle to their eyes, the insects end up flying in circles, closer and closer to the light.

Nighttime Activity

Find some night fliers

Grab a flashlight and get a close-up look at some night-flying insects.

You'll need:
- an outside light or a flashlight
- a magnifying glass (optional)
- a field guide to insects (optional)

1. Turn on an outside light or flashlight on a dark summer's night. Check out the flying insects that are attracted to the light.

2. Use a field guide to help you identify moths, lacewings, crane flies, caddisflies, June beetles, mosquitoes and other night fliers. How many different kinds can you count?

3. Compare the sizes, shapes and colors of the insects. Use a magnifying glass to get a close look at insects that land near the light. Notice the kinds of wings that different insects have — scaly, clear or hard.

Gypsy Moth

Lacewing

Crane Fly

Find some night crawlers

Slugs, snails, earthworms and other moist-skinned creatures hide during the day to protect their bodies from drying out. On warm nights they slide out on their slimy trails to search for food. Get a close-up look at some night crawlers with this simple trick.

You'll need:
- a flashlight
- a piece of red tissue paper
- an elastic

1. Cover the light of a flashlight with red tissue paper and fasten it on with an elastic. Now your light will appear red. Since night crawlers cannot see red light, they won't try to avoid it.

2. Go outside with an adult at night to look for night crawlers. Look for earthworms by shining your flashlight on damp soil, especially in gardens.

3. Check out low plants in gardens, woods and other damp places for slugs and snails. Other night crawlers to look for include sowbugs, walkingsticks, ground beetles, spiders and daddy longlegs.

Earthworm

Daddy longlegs

Sowbug

Garden Snail

Ground beetle

Walkingstick

Night prowlers

Have you ever heard the crash of a garbage can being knocked over by raccoons or the howl of distant wolves in the night? Whether you live in the city or the country, you may hear the sounds of animals that come out after dark.

Where do nocturnal animals go during the day? Many night prowlers in the city spend their days in cemeteries, where there are often lots of trees and bushes around for shelter. There aren't many people to bother them there, either!

Raccoons leave their dens in hollow trees or an attic to rummage through garbage cans and gardens for food. Rats and mice scamper through the dark looking for scraps of food and trying to avoid the cats and dogs that are also out at night. Bats hunt insects in the air with their high-pitched squeaks, too high for humans to hear. And skunks root under stones or in lawns for tasty grubs and other insects to eat. If skunks are disturbed, they spray their strong odor into the air to keep predators (and people) away.

If you live in the country, or visit a cottage or campsite there, you may hear wolves or coyotes at night. Coyotes, which are more common near people, look for food around farms in the night. Wolves try to avoid people, and you'll hear them more often in the wilderness. Wolves howl to draw the pack (group of wolves) together before hunting, or to defend their territory from other wolves. Listen for wolves howling after dark in late summer when the adults are teaching their pups to hunt and howl.

Badgers, opossums and mink move around quietly after dusk in search of food. These shy creatures will quickly find shelter if they are frightened. Flying squirrels glide through the air using the broad flaps of skin between their outstretched legs as miniparachutes. If you have a bird feeder, you might see a flying squirrel coming in for a landing after dusk to feed on seeds.

Some zoos have exhibits of nocturnal animals. Since you can't visit them at night, they are kept in special buildings lit during the day with red lights. Because nocturnal animals can't see red light, they think it is dark and remain active. Brighter lights are used at night so the animals will sleep.

Badger

Night at the shore

You may have been to a beach during the day, but have you ever visited one at night? Instead of people and beach balls, you might see some interesting wildlife. Night at the shore can be a busy place. Here's an easy way to take a peek at what goes on underwater at night.

You'll need:
- a flashlight
- a heavy stone
- a clear jar with a tight-fitting lid, big enough to hold the flashlight and stone
- some string or rope

1. Visit a pond, marsh or seashore with an adult at night.

2. Turn on the flashlight and put it in the jar with a heavy stone.

3. Tightly seal the jar with the lid so no water will get in.

4. Tie the string firmly around the neck of the jar and hold onto the other end of the string. Carefully stand on the shore, or kneel on a dock or boardwalk and lower the jar slowly into the shallow water. What can you see in the water below?

Here's what to look for:

- Crabs, lobsters and shrimp are active at night, feeding and moving around on the sea floor.
- Tiny floating animals called zooplankton rise up near the surface at night, when they can feed more safely than during the day. Although the zooplankton are too small to see, you may see fish and other aquatic life that eat them.

• Aquatic insects such as diving beetles, water boatmen, backswimmers, whirligig beetles and others may be swimming around in ponds and marshes.
• Crayfish move along the bottom of ponds and streams.

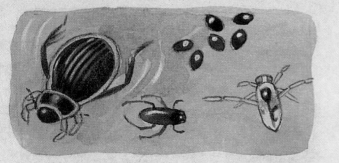

Shore watch
There's lots of activity on the shore, too. You may catch a glimpse of a mink or beaver looking for food, or a deer or moose having a drink at the water's edge. Many birds that live near water are active at dusk or through the night. Night herons, bitterns and rails prefer to feed in the twilight in marshes and along shorelines.

Turtle time

At certain times of year, thousands of tropical sea turtles climb onto their traditional nesting beaches at night, where they dig holes in the sand and lay their eggs in the dark. Weeks later, the young turtles hatch at night and make their way across the sand to begin their lives in the sea. Although darkness helps the tiny turtles hide from predators, many are eaten by nighttime hunters such as birds and crabs.

A desert night

Imagine living in a place that is so hot and dry during the day that you'd have to hide from the Sun to survive. That's what life in the desert is like for many animals. To stay cool, these animals sleep during the day and become active at night. Look below to discover how desert animals are adapted to life at night. Then take a look at their daytime hiding places on the next page.

An Elf Owl gets water as well as food by eating insects and spiders. A spider's body is more than 80 percent liquid.

The Sidewinder is a kind of rattlesnake. It hunts in the dark using heat-sensitive organs called pits. They help the snake find mice and other warm-blooded prey.

A kangaroo rat has very good hearing so it can avoid night predators such as Kit Foxes and snakes. Its large hind legs and feet are perfect for hopping quickly, like a miniature kangaroo, to escape danger.

The Banded Gecko has large eyes that open wide to let in as much light as possible so it can find food in the dark. To protect its eyes from light during the day, the lizard's pupils close up to tiny holes.

The Kit Fox's long ears can move around to listen for the scurrying sounds of mice and other rodents.

Desert plants at night

Many desert plants bloom at night, instead of in the day, to attract nighttime pollinators. For example, the Saguaro cactus is pollinated by Longnose Bats, and yucca flowers are pollinated by Yucca Moths.

Night singers

You've probably heard lots of birds singing during the day, but did you know that some birds also sing at night? Owls, nighthawks and Whippoorwills sometimes call in the dark. Ovenbirds sing in the woods, bitterns boom in the marsh, and robins, other thrushes and blackbirds occasionally sing at night if they've been disturbed. Birds aren't the only animals to fill the night air with song.

Have you ever heard a ching ching sound, like distant sleigh bells, coming from a wetland on a warm spring evening? It's a chorus of tiny treefrogs called Spring Peepers singing to attract mates. They may be joined by a variety of other frogs and toads, each with its own tune. Listen for the banjo-like twang of common Green Frogs and the deep chug-o-rum of Bullfrogs. These creatures make their sounds by forcing air back and forth from their lungs to the bubblelike vocal sac under their chin. As the air passes through their throat, it vibrates the animals' vocal cords and makes the sound you hear.

Bullfrog

Spring Peeper

A living thermometer

On warm summer evenings you may go to sleep listening to the chirping of crickets. The common field cricket also sings during the day, but its voice is often drowned out by other daytime noises. The males sing to attract a mate by rubbing the sharp edge of one front wing over a ridge on the other front wing.

Scientists have discovered that the chirps of the Snowy Tree Cricket can help you tell the temperature. This works best on a warm summer evening. All you need is a watch or clock with a second hand, and a paper and pencil.

1. Count the number of chirps you hear in eight seconds and add four. The answer gives you the approximate temperature in Celsius.
2. For Fahrenheit, count the number of chirps in 15 seconds and add 37.

Snowy Tree Cricket

Green Frog

Plants at night

If you tour a flower garden on a sunny day, you may see beetles, bees and other insects visiting flowers to feed on them. At the same time, they carry pollen from one flower to another, called pollination. Flowers must be pollinated before they can make seeds. If you return to the garden as the Sun sets, you'll see that many brightly colored flowers close up their petals to keep their pollen dry during the damp night. Sometimes the daytime-pollinating bugs are trapped inside the closed flowers all night.

Other plants stay open day and night or are closed during the day and open up at dusk to attract night pollinators such as moths, fireflies and springtails. Take a look at these night-pollinated plants. Here are some things to notice:

• They are white or pale colors so they can be seen more easily in the dark.

• Their petals are often long and narrow, and notched or divided so they show up well against the dark background.
• Flowers pollinated by moths are often tube shaped so that only the long tongues of moths can reach the sweet nectar.

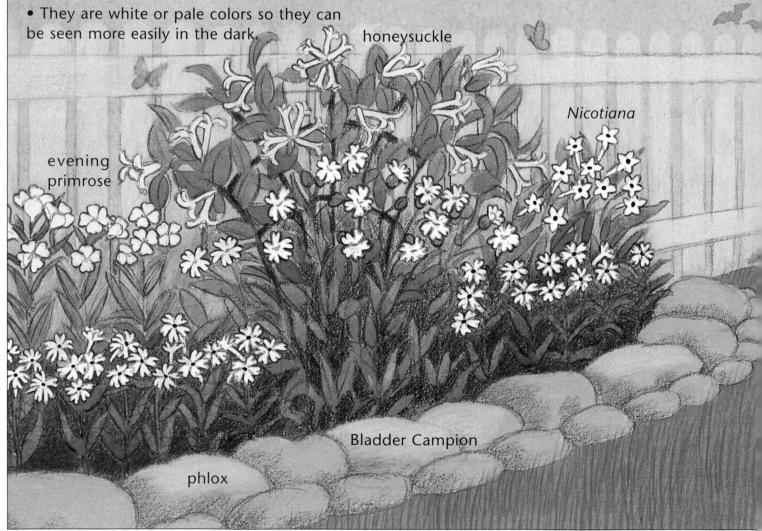

honeysuckle

Nicotiana

evening
primrose

Bladder Campion

phlox

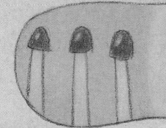

Fungi grow best in the damp, dark conditions of night. A stinkhorn fungus can grow up to 20 cm (8 in.) overnight!

Strong scents also attract pollinators, so many flowers give off perfume at night. Not all night flowers smell sweet, though. In the tropics, nectar-feeding bats are attracted to the wild-banana flower by its stink.

Sleeping flowers

Try to find some of these flowers in local gardens or fields and compare what they look like on a sunny day to how they appear after the Sun has set.

hawkweed

dandelions

morning glories

daylilies

four-o'clocks

tulips

crocuses

Growing in the dark

You can compare how much plants grow during the day and night by growing your own plants indoors.

You'll need:
- corn or bean seeds
- two flowerpots
- potting soil
- water
- measuring cup
- ruler with millimeters marked on it
- paper and pencil

1. Fill each flowerpot with soil. Poke a hole about 1 cm (¹/₂ in.) deep in the soil and plant one seed in each pot. Cover the seed with soil. Number the pots and keep the soil damp. You will observe both pots in this experiment, just in case one plant doesn't do well.

2. When the plants, called seedlings, appear, place each pot in the same window. Water each plant with about 50 mL (¹/₄ c.) of water every three or four days.

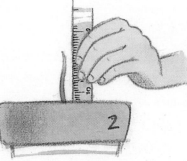

3. Measure the plants every 12 hours. At the same time each morning, use a ruler to measure the height of each plant. When you are measuring, place one end of the ruler on the surface of the soil and measure to the tip of the plant's stem. Measure the plants again 12 hours later, before you go to bed.

		Pot 1	Pot 2
Mon.	a.m.		
	p.m.		
Tues.	a.m.		
	p.m.		
Wed.	a.m.		
	p.m.		

4. Make a chart and write the morning (a.m.) and night (p.m.) measurement for each day beside each pot number.

5. At the end of a week, figure out how much each plant grew every day and every night using these calculations:

height of plant at night
− height on same morning
_____
= daily growth

height of plant in morning
− height on previous night
_____
= nightly growth

What's happening?
Plants do most of their growing in the middle of the night when you're asleep. That's because the substance that makes plants grow, called auxin, works best in the dark.

Lean into the light

Have you noticed that your houseplants always grow toward the window? During the day plants grow and bend toward the light. This is because there is more auxin on the shady side of a plant. The shady side of the plant grows faster, making the plant bend to the light. At night some plants bend toward the moonlight for the same reason.

Lighten up

Even though plants keep growing at night, they need daylight to stay healthy. In this experiment you'll find out what happens when a plant is put in permanent darkness.

You'll need:
- two small potted plants that are the same kind, such as geraniums
- a cardboard box with no holes
- water

1. Place one plant in a window and put the other in a dark room with a cardboard box over it so no light gets in. Water both plants equally.

2. After a week, compare how the two plants look. Compare them again after two weeks. Take your plant out of the dark before it dies and nurse it back to health.

What's happening?

The plant in the dark will start to weaken after a week or so. You may notice leaves turning yellow, dying and falling off. If the plant is left in the dark too long, it will die. This is because plants need water, air and sunlight to make food. When it is sunny, green plants take in carbon dioxide from the air and water from the soil to make sugars that the plant uses for food.

This is called photosynthesis. At night, without the Sun, no photosynthesis occurs.

Plant respiration, or breathing, goes on day and night, though. Respiration is the opposite of photosynthesis. A plant takes in oxygen from the air and combines it with the stored sugars in the plant to produce energy for plant growth. At the same time, carbon dioxide is released back into the air from the leaves.

Photosynthesis occurs during the day

sunlight
+ water from the soil
+ carbon dioxide from the air

= sugars stored in the plant
+ oxygen and water released back into the air by the leaves

Respiration occurs day and night

oxygen from the air
+ sugars stored in the plant

= energy for plant growth
+ carbon dioxide released back into the air by the leaves

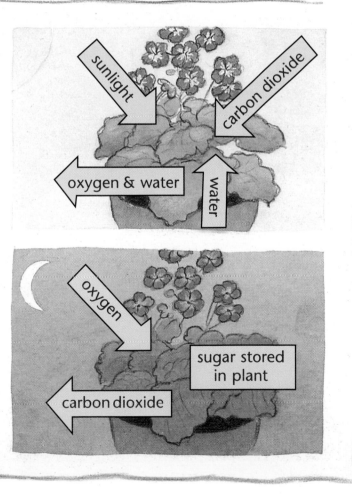

Night migration

As the days get shorter and colder in fall, millions of birds head south to find food. You may see large flocks of geese, hawks or blackbirds migrating during the day, but many birds also migrate at night.

Migrating in the dark helps small birds hide from predators. It also gives insect-eating birds, such as flycatchers and warblers, time to catch food during the day when they can see. Food gives them the energy for their long flights at night. The cooler temperatures and higher humidity at night make it easier for birds to cross above hot deserts on their migration routes.

You know how hard it is to get around in the dark, so how do migrating birds find their way at night? Many birds use the stars as a guide to fly south. Migrating birds usually fly higher at night than in the day, to stay above storm clouds and keep a clear view of the stars. In experiments done with caged birds in a planetarium, researchers found that all the birds faced south when they were in a room where a normal pattern of stars was projected on the ceiling. When the star pattern was turned around, the birds also turned around. When there was no star pattern, the birds faced in all different directions.

Since many birds migrate at night, it's difficult to see them. To keep track of where they are, scientists use several techniques, including radar, radio tracking and bird banding. Some people use binoculars or telescopes to watch for migrating birds crossing in front of the Moon on clear evenings in spring and fall.

Toad crossing

Toads also migrate at night. You may see a large procession of toads crossing a road on a warm spring night. The toads are heading to their breeding ponds.

Index